–A– DRIFTERMAN's –DIARY–

by
J. E. Holmes

ACKNOWLEDGEMENTS

I would like to thank the many people who helped me during the course of putting this book together including the Lowestoft Maritime Museum, Lowestoft Library, David Butcher for kindly writing the foreword, Keith Skipper for our dinner-time chats, Dean Parkin for his advice and Christine Johnson for proof-reading the manuscript.

Special thanks go to the George family who shared with me their memories and allowed me to record them for future generations.

© J. E. Holmes 1994

All rights reserved. No part of this publication may be reproduced, stored in a retrieval system, or transmitted in any form or by any means, electronic, mechanical, photocopying, recording or otherwise, without the prior permission of the publishers.

CONDITIONS OF SALE

This book is sold subject to the condition that it shall not, by way of trade or otherwise, be lent, re-sold, hired out, or otherwise circulated, without the publisher's prior consent, in any form of binding or cover other than that in which it is published and without a similar condition including this condition being imposed on the subsequent purchaser.

Typeset by Chemtech Graphics
Sussex Road, Gorleston, Norfolk

Printed in England by Sellin' Print, Lowestoft

ISBN 1 872992 98 6

Foreword

by David Butcher

I feel honoured to have been asked by Jim Holmes to make some introducing comments regarding "A Drifterman's Diary". The material he collected was valuable at the time, but has become increasingly so as the twentieth century draws to a close. Herring fishing was once of great importance in the local economy of the East Anglian coastal fringe and this is reflected in the pages which follow. But there is more to the book than mere financial significance. Life in fishing communities was very different from that in agricultural or industrial ones, and aspects of the maritime culture are plain to see as one reads through the reminiscences of Jim's respondents. All of them were born into fishing families and all of them reveal something of the tough and demanding lifestyle which was their inheritance. In the first of his oral histories, George Ewart Evans used a line from an old Chinese poem to provide the title: "Ask the Fellows Who Cut the Hay". That, of course, is always the best thing to do - get information from people who have done the job, whatever it is. Jim Holmes was well placed to record the remembrances of Norfolk fishermen and their womenfolk, and it is good that something of their collective experience should find expression in this book.

David Butcher
September 1994.

Gutting the herring. The fisher girls usually worked in crews of three with three baskets behind them. As they worked they sorted the fish, throwing them in the baskets behind them according to size, without even glancing over their shoulders!

Introduction

When I was a boy, Yarmouth was dominated by the herring industry. I never joined the fishing although I did go to sea on drifters as a passenger and my uncle Sam Woodger was famous for the widely known 'John Woodger and Sons Kippers', popular in dining cars of railways, Atlantic liners and hotels everywhere. John Woodger came from the Newcastle area and introduced the north country method of curing or kippering herring to Great Yarmouth, and his sons greatly expanded the business which was later taken over by MacFisheries (Leverhume).

Early in the year we used to see certain boats preparing to go to Plymouth and Penzance, the 'westward' as they called it. They had been surveyed, overhauled, fitted out, repainted, coaled and provisioned and a pretty sight they made steaming out for trials and compass boxing. A very important man was the compass adjuster. How I longed to go with them but never dared ask although I'm sure they would have taken me for that short trip as fishermen were always kind to boys.

Later some boats sailed northward for the Scotch fishing. Many a time I've watched them steaming past Caister, Scratby and Winterton where some of the wives and sweethearts gave tearful little waves and the drifters hooted as they steamed away to fish the Minch, the Hebrides, Pentland Firth and other parts, gradually working their way home via Wick, Fraserburgh, Aberdeen, Shields, Scarborough, Grimsby and Dogger Bank then back for the autumn fishing or 'home fishing', the main voyage of the year. This was the climax of the year's fishing and hundreds of English and particularly Scottish drifters converged on Yarmouth and Lowestoft crowding the quays.

The Scots boats always lay in on Sundays but English boats went to sea and so the Scots moored very closely, especially near the Town Hall in Yarmouth, while the English boats lay further down-stream so as to get to the sea unhindered. The drifters lay thick in the river and they say it was possible to cross the river from boat to boat!

Had anyone in 1950 told us that in about 20 years there would not be a single herring drifter working out of Yarmouth or Lowestoft he would have been laughed to scorn. Few today can imagine the vast quantities of herring landed. Herring, herring everywhere. Gulls screaming, boats hooting as they nosed into the crowded quays or as they came astern. Monster piles of barrels near the pickling plots and yards, hundreds of crews of Scots girls gutting and packing like lightning at the troughs, out in the open in all weathers, brisk and hardy. Whilst waiting for the boats to return they often linked arms and paraded the roads gaily singing their Gaelic songs also, in such spare time as arose, out came their knitting. It was marvellous to behold how without looking at the work, talking all the time, they knitted elaborate designs without using patterns. It was in their heads!

What an animated scene: Scots girls, coopers, carters, curers, salesmen, shipping grocers, butchers, drifters crowding the quays, everywhere a hive of activity, the like of which we'll never see again but will remember as long as we live.

Dennis George of Winterton was born in 1877 and epitomises the best of the breed. His life spans the great days of the fishing from sailing luggers to steam drifters as fisherman, skipper, boat owner and recognised authority on herring and herring catching. He speaks for them all. When I first knew him he was over ninety years of age and although in bed following a severe fall, he was as bright as a bee and I have faithfully recorded his experiences of fishing. I've always liked fisherfolk and have a great admiration and respect for this wonderful breed. Fishermen are a breed apart, hardy, independent, skilled in their occupation, fine seamen, the backbone of England, second to none.

Jim Holmes
November 1994

Chapter One:
A YOUNG FISHERMAN

Dennis George was just ten years old in 1887 when his father first took him to sea. He remembered, ".... The Winterton schoolmaster had no more to teach me and so put me to work in his garden. My dad said to me, 'How are you getting on at school?' and I told him that I'd been put in the garden. 'Oh well,' he replied, 'I'll take you to sea with me in July.'

"I went to sea in July on the *Christiana* who belonged to old Durrant of Yarmouth. Her number was 500 and she was a sailing lugger. That was before they had a steam capstan. They had a crank and some had two bars go across the capstan so they go round and round to heave your ropes in and get the nets alongside. In that time we had about 12 warps about 120 fathoms each and you left one with a swing when you shot your nets. When wintertime came my Dad said, 'I aren't going to take you to sea in the bad weather,' so at the age of ten I had to find something else to do.....

"My uncle said, 'I can do with you. You can go round and sell the fish for me.' Twice a week with a horse and cart. You start at Somerton, then Horsey, Martham, Rollesby and then head home through Hemsby.' I sold red herrings, bloaters, shrimps, crabs, flat fish and whatever he bought. The best red herring were two for 1½d and bloaters were ½d and ¾d each, shrimps were 2d pint ready boiled....."

This was in 1887 but in the following year Dennis's father took him to sea again in the *Adviser* (No.499). Dennis explained, "....She belonged to my uncle. He and his brother had two ships each. Then in the autumn my father wouldn't take me to sea again in the wintertime. My uncle had sold the pony and trap so this time I found a job on Empson's farm. I went to work for him at 2/6d a week. I had eight cows to feed, fetch turnips and mangels, feed the pigs and the bull and cut hay for four horses although I didn't have to feed them. I was eleven years old and I started at six-thirty in the morning, had an hour for breakfast, an hour for dinner and went home at half past six. I had to come back at eight until eight-thirty to give some hay to the cows and bull and then go home to bed so as to be ready for six-thirty next morning....."

Dennis worked there for five years until he was sixteen and by that time was paid 4/6d a week. When he was sixteen he went to Lowestoft and got a berth on the *Young John*. Dennis remembered, "....She was only 28 tons and belonged to John Jenner, a very nice man. After we'd been away a fortnight working about forty miles out from St. Ives, when the ship rolled I couldn't make water so they had to put me ashore. Well, I was ashore for a fortnight and my water came all right and the next trip my water was A1 and I never had no more trouble with it. In the westward we didn't fish Saturdays and Sundays, we did as they did down there."

When Dennis returned from the westward his skipper, John Jenner, decided to stay at home and work in the beating chamber he had at Pakefield. Jenner offered him a place to live and work in the chamber and so Dennis went to stay in Kirkley with Jenner and his wife, who was a laundress. Dennis continued, ".....We used to walk to Pakefield. At the net chamber the ropes were coiled outside and the women used to beat the nets in the lower chamber, while men used to ransack the nets in the top chamber.

"In the autumn David Jermany became skipper of the *Young John* and I went to sea until Christmas. I started as cook. In the following year, 1895, I was eighteen. My sister had a boy die with diptheria and my other sister Chris Emma and my mother went to see her and caught it too. I only spoke to them at the chamber door and I caught it as well! My father had to stop ashore to look after us three as we lay a-bed and gargled our throats with vinegar and water and some kind of Condy's fluid. My sister and my mother got better but mine wouldn't give way so it had to be burnt off with caustic soda. My throat was sore for years after that but I still went to sea. Then I had tonsillitis and instead of having them taken out, I had to stand against a window and the doctor took a knife and slashed the tonsils without any anaesthetic. In a few years they got all right but it was six months before I could do anything.

"In July I went in the *Olivette* belonging to Pitchers of Yarmouth along with Fred Goffin, a Winterton man, who was skipper. We had no steam capstan and we got caught in a gale on the North Sea and spoilt half a fleet of nets. We were shot 2 or 3 miles off Smiths Knoll when it came on to blow a gale of wind and we had half the nets in but we couldn't get anymore. We had to hang on to the other half but they were all spoilt and we lost them. We were up to the Gasket Lightship before we got the ropes in so we'd drifted a long way.

"I stayed with the *Olivette* until Christmas and then I got a berth in the *Nellie Jane* (No. 344). Jimmy Smith, the owner's son, was skipper and we went through the whole of that year (1896) and never got caught in a gale of wind. Mackerel fishing, North Sea, home voyage - whenever there was a gale of wind we escaped it!

"In the following year Jimmy Smith had a new boat. Her name was the *Forward* (No.708) and we went away to fish off Scarborough, everything new. The mate had the watch. With a strong SW wind and rain, the ropes wouldn't go through the blocks, they were all brand new and swollen by the sea water. We used to have a little compass in the binnacle, it was on a pin and used to swing round, not like these ones in oil. We got into the Humber and we got onto the Stonebanks. We nearly lost the ship, we should have been five or six miles north of Flamborough Head, in view of Scarborough. The waves kept rocking the boat and we were wet through. As we went into dock the next morning there was a sailing smack going ahead of us and they were singing out to us on the pier, 'Get your sails in whatever you do,' but we couldn't budge them, the ropes were so swollen. Our mizzen sheet caught on a capstan on the pier head and brought us up else we should have sunk this here smack at the dock gates!

"We went out of there and worked off Scarborough. We got 12 miles E by E off Flamborough Head when down came a heavy gale of wind and we set the balanced reef sail, that's four reefs in the mainsail and four reefs in the mizzen and a storm jib. We used to take two reefs in the jib so it was a storm foresail and put a small jib on the bowsprit.

"We shipped a heavy sea. We were using the yellow dishes and pots then, not the sort which didn't break, and so everything was smashed up and we hadn't anything to drink out of. So we took our sou'westers and turned them inside out and drank tea out of them! The wind blew very hard and somehow we managed to get to Lowestoft.

"The following year I was in the same boat. We started on St. Valentine's Day and we went to sea on the 19th and went away on the Tuesday morning to fish for mackerel at Plymouth. When we got to Southwold the wind kept freshening and freshening and we had to keep taking a reef in, then we had to take two in until we got down to the balance reef sail and that night we could see the North Goodwin. Some of the boats went into Margate, lost their bowsprits and went and brought up in Margate Roads. We kept the North Goodwin in sight and kept tacking and kept in sight. Well, next morning we'd got into the cod channel, that's a channel inside the Goodwin Sands up towards Dover and Margate and we beat a bit through there and got into Dover. Out of Dover we went into Newhaven and sheltered there for a little while then we went on to Plymouth." This was a hard rough voyage compared with the fishing lugger *Paymaster* which on February 2nd, 1860 sailed the 240 miles from Yarmouth to Portsmouth in 23 hours.

Dennis George ran into still more gales off Plymouth. "There were three of us going out of Plymouth, thirty miles out of the Sound and about twelve or fourteen miles outside the Eddystone Lighthouse. All three of us got about a mile or so apart because if you shot a mile of nets you had to be a mile apart so you didn't swing into one another if the wind shifted. We all of us started to shoot, *Our Girls*, the *Violet* of Lowestoft, a brand new boat that never shot a net before. After we'd shot about 25 nets, our skipper said, 'It's going to blow a gale of wind, we'll haul ours back again.' Well, we turned round and hauled those 25 nets back again. It was blowing a gale of wind and that night the *Violet* was lost with all hands and *Our Girls* lost everything, spoilt and lost all his nets but survived and got through it. The skipper of the *Violet* was called Ayres and was brother to Mason Ayres who used to have the boats at Lowesoft. That was in 1898.

"In 1899 I went into the *Our Girls* along with George Gladwyn and the gale of wind came down on 6th April. There were seventy of us shot about seventy to seventy-five miles NNW of the Longships. Our little boat did everything except turn over when we were adrift. We didn't part from our nets until five-thirty that morning, most of them parted late that night."

Further troubles added to the fishermen's difficulties though, "We used to go eighty or ninety miles from the Longships and one hundred miles from the Scilly Isles and sometimes only forty miles. Once I remember we got a shot only fifteen miles from the Longships. We got a few herring but they were stocky bait. Stocky were the fish, other than mackerel, sold to fishermen for bait. A last of mackerel was ten thousand fish. At that time we would tell up the catch (count them by hand). For mackerel 120 to the 100 (long hundred) and for herring 132. The number of fish over 100 would go to the fish salesman. Years ago we used to throw pieces to one side for the stocky bait for the crew. When I first went in the sailing boats the skipper used to get $1^{3}/_{4}$ shares and the cook $^{1}/_{2}$ share and so on. In the steam boats it was different but fairer.

"I was coming home in the month of June and we came across what we call the 'muldoonas', they used to come from the Spanish coast and swim with their fins up like an umbrella. They'd spoil your nets, it was like slime and after three weeks the nets would be as rotten as a pear. We once lost thirty nets in four warps. They used to come in June and we had to leave, they were right big fish. They'd go from the Channel and in the steam boat times we'd meet them off Lerwick in September. We also used to get sharks in the month of June. I remember when I was in the *Forward* we shot about twenty miles from the Scilly Isles and we got five great sharks, not the bullnose ones but the maneaters. We got them out of the nets and they lay with mouths open across the thwarts boards. I took a rail about twelve feet long and stuck that right down the throat of one, he clamped his teeth and they met. He'd bite your legs off if he'd got anywhere near you then he laid there until he died and we heaved him overboard. We had to leave if they were around."

A steam drifter just putting the sail up around the beginning of the century. This was the sort of low-powered boat that Dennis George would have served on in the early days.

Fishermen 'telling them up' by hand, counting the fish into long hundreds. The man in the hat was probably the chief salesman or owner judging by how well-dressed he is.

At sea there is constant danger but fishermen accepted the dangers and difficulties of their calling and took it all in their stride. Many sailing boats were lost and no-one would know what had happened. In his matter of fact way Dennis George told me of one near disaster that might well have had tragic consequences. By quick thinking and desperate effort, but mostly by the luck of having air trapped in the sails which prevented the vessel completely turning turtle, they righted the ship and lived to tell the tale. Many crews were not so lucky.

"....I was in the *Star of Peace* again along with George Gladwyn. We went out from Newlyn fifty miles NW of the Longships and at three or four o'clock it came down a heavy gale of wind. We had set the balanced reef sail and we shipped a heavy sea and it cleared the decks, took the foresail away, threw the little boat into the mainsail and threw the iron ballast from the weatherside to the portside and she lay over with the mainsail and the mizzen on the water which stopped her from going right over.

"We'd lost the swipes, thwartsboards and everything. The wind eased and it seemed almost like a calm. We got the little boat out of the mainsail and lashed it upside down so she didn't fill with water. We were battened down tight but took a little hatch off and three of us went down below and put the ballast back, half hundredweight iron pieces with two little holes in each end. We put that back in the starboard side, shovelled the shingle back again and righted her on the keel. That blowed a gale of wind from the Monday to the Friday and we lay about all that time with a balance reef sail.

"Friday afternoon about three o'clock the wind eased though there was a lot of swell. The skipper said, 'We'll try and shoot the nets.' We had the mast down and the chain out and we were a long while before we could get her up before the wind. We had to shackle the chain out and then we put two bags over and lashed two trunks with a rope in the middle and put them over so they were full of water so we didn't run and could shoot the nets easier. The mate put a lashing around his waist and the skipper stood at the tiller and took a great lump of water which knocked him away from the tiller and soaked him through, I even got wet in the kitty shooting the nets. They were using a sea anchor.

"It fined away the next morning and we got a last of mackerel. We took them to Newlyn and they made £100...."

Chapter Two:
DAYS OF SAIL

Dennis George had no illusions or sentimental nostalgia for the days of sail. "....It was awful in the sailing boats..." he said. It was indeed, very hard work and dangerous. ".....In the sailing boats we didn't have a fo'castle, we had a cabin. When I was in the *Nellie Jane* we always used to have two on a watch. The bunks weren't long enough so we had to knock the bulkhead down so a man could lay his head one way and his feet the other so our feet used to be close to one another, packed like sardines. We had nine in a crew for herring, eight after mackerel, sometimes even seven. I once went net stower all alone but it was a heavy job.

"I went cook along of John Jenner in 1894 in the *Young John*. She was only 28 tons and had a little cabin and when we sat down in there one of the mates was six feet tall and his boots used to go right across the cabin to the after end! When we sat down we had to sit legs between each other and get our meals. It was only wide enough in the fore part of the cabin for two people to eat. Plenty of them lay on the lockers and they lay three in one bunk, three of us younkers had to lay in one bunk. I used to wait until the other two got in then I used to lay on top of them.

"There was no table. We were stowed up for room. There were bunks but not big enough to sit in. For cooking we had a stove down in the cabin, an open stove. We used to have a cooking tin with a lid to keep the cinders off. It had a handle and we used to put it underneath (in the ashes under the grate) and stir the coal up and that used to cook from the top. We cooked everything like that and then we put the boiler on. We used to peel the potatoes and put them in. We used to have a lot of sweet duffs boiled in a cloth and when we took them out there used to be a thin layer of skin due to the salt water. When you peeled that off you would eat it and that would taste lovely..."

When I suggested fishermen lived well on the drifters he agreed but only if they had a good cook. On the sailing boats it was a different story though, "When I first went on a sailing boat we were allowed two tins of milk a week, one and a half stone of sugar, two pounds of tea which had to do 21 kettles, a kettle for breakfast, a kettle for tea. We used to put the sugar, milk, tea, everything in the kettle. We always had a kettle when we hauled the nets at night time.

"We had about 3lbs of beef a day all for nine people. When I got to be skipper I said, 'I'm going to feed my chaps well, if they don't feed well they can't work well.' We used to have a dozen tins of milk and four or five pounds of tea. We used to have bacon and eggs for breakfast on a Sunday morning and I used to get a leg of mutton or a leg of pork and we used to cook that for our Sunday dinners and we used to have dumplings beside. By then we had a proper cooking stove in the sailing boats. Later in the steam boats we had a galley, that was a big improvement. There was no lavatory though. We had to sit over the side, you'd go to leeward and catch hold of the rail or rigging, there was no bucket in those days. We didn't have fresh water to wash in, you had to rinse yourself in salt water. If you were away eight weeks you'd come home as black as a crow. I've known my father to be away for eight weeks and he would come home as black as a tinker. He used to sit down, take off all his things and my mother would wash him all over and get him clean again!

"It was awful in the sailing boats. We didn't have much fresh water. We used to have two barrels on deck with a piece of canvas nailed over and a cork put in. When you drew water you had to be careful not to get any salt water in. You had to watch out! I remember in 1895 when we used to have the salt locker down the cabin to salt the fish with.

"We used to have fourteen tons of iron ballast in a boat of forty tons and ten tons of shingle on top of that. I was in the *Forward* when she was new in 1897, she was built on Chamber's Yard at Lowestoft and belonged to Jimmy Smith of Lowestoft. She had all American sails. They were heavy built, our spinnaker was 265 square yards and the big top sails when we had them in with the full sails and the stay sails up covered 1,100 square yards of canvas. Then we used to have a bonnet on the mainsail....."

The sailing luggers had to adjust their sails according to the wind. Their lives and the safety of the vessel depended on their sound judgement.

"....When the wind was about force three or four we used to take the big topsails down and set what we called the jack topsails, take the spinnaker in and put the eleven course jib out and set the to'foresail. We used to take the bonnet off the mainsail when the wind was another point higher. The bonnet was made with holes all the way along so you threaded the lashing through and put it on like a butt so you could take the two reefs out of the way and didn't have to roll it up. When the wind was about force five we used to take the bottoms off the mainsail and the mizzen, take the big jib in and set the seven course jib and the stay foresail. Take the to'foresail in. After that if the wind increased we used to take another reef in the mainsail and mizzen and if it increased a little more we used to come down to the balanced reef sail and set the storm jib, five course jib, reef the stay

Above, the fleet leaving the port with Yarmouth harbour on the right and Gorleston Pier on the left. This was in the heyday of the fishing, 1913-36, when these vessels used to come out in their hundreds. Below, the fleet can be seen returning to harbour. These are mostly Scottish boats.

PHOTOS: Mr. P. Trett

foresail and we were fixed for a gale of wind. The smaller luggers only used to set five to six hundred square yards, they were about 28-30 tons. Later we had to do away with our big spinnaker. You could stand on the after deck and hold the sheet in your hand but they were too heavy so they did away with it and made stay foresails for the trawlers and made light ones, two hundred and forty-five square yards, that was plenty big enough. We used to ship the bowsprit out, there was a chain from the foot of the stem to the end of the bowsprit. The mainmast was about forty feet. I've been running through the Runstern Gap from the Longships towards Newlyn after a heavy gale and there were forty foot seas, they didn't used to break. I've seen two or three sailing boats running with us drop into the trough of the waves till we lost sight of them then up they came on the top of the waves all right. They were big Atlantic rollers, that was a sight that was...."

Fine seamen as they were, disaster was ever near. ".....My grandfather was drowned off Goodwin Sands in 1845. He was master along with Bob King's father in a little sailing lugger. It was lost with all hands going to fish off Plymouth in January of that year...." Sailing boats depended on the weather. "....Sometimes we'd have a head wind and I've been a beating all the way up the Channel to Dover and thought we'd get a fair wind when we got around Dover only to get a NE wind and have to beat all the way home. Sometimes you had a fair wind and it didn't take you long, it was like that in the sailing boats."

Chapter Three:
THE BEATSTER'S LIFE

Every owner had his large or small net store, a chamber for beatsters and ransackers. There he kept all the things he needed for his boat or boats. Dennis George's daughters, now Mrs Larner and Mrs. Wodehouse, shared with me many memories of the fishing community of yesteryear. ".....I first went to school when I was about two and a half. We used to go down to what we called the reading room. We just had sand on slates and take a finger and just draw shapes and that sort of thing. There were about twelve of my age and other children. The schoolmaster's wife used to take us until we were five then we went up to the big school. It was all free schooling. You see, people had such big families that time of the day and the mothers were glad of children to go to school as they hadn't time to look after the children at home.

"I left school when I was thirteen and started in the beating chamber when I was fifteen. My mother taught me and she had six more apprentices besides me. We used to go six months for nothing and then six months for 2/6 a week, six months for 5/-, then after three months we got 7/6 and that was our lot. It took three years to learn the trade. It was a two storey building, we were some down in the bottom and some in the top. Twelve altogether in North Market Road (it's a factory now).

"I worked for my father-in-law although he wasn't at that time. He was a local boat owner and had the *Charm*, the *Queen of the Fleet*, the *Achievable* and one other. The *Charm* was a very small boat and was one of the very first steam drifters. I forget how long the nets were. On the side wall we used to have a cleat, that was a big hook, and we gathered the net in our hands and put a piece of string round and hook it up on this cleat, get about six yards from the cleat and then what one called 'tricklin' the net over and see where the holes were. When we come to a hole we had our needle and cotton and then we'd start mending the hole with the cotton. There was all the ropework, we had twine and if the sidelines were gone we had to twist the cotton in a special way all down the side to strengthen it and on the end of the nets there were big thick ropes about one inch in diameter. Sometimes they got broken and that was the men's work. That was called the 'heading' and that was very heavy work, it would tear your hands to pieces. Sometimes you'd got a net and there wouldn't be two holes in it, then you'd get another net and the whole middle would be ripped out and you'd have to get a piece of lint from old nets and what we called shoot it all in. We used to use soap when I first went to make the cotton pliable but after that we needed to use linseed oil. When the boats came away from Shields they brought the nets home and they used to be full of jellyfish. They'd all gone to powder and if we were trickling the nets to find the holes, this powder all went up your nose and in your eyes, they used to stream, it was nasty, and it was very uncomfortable.

"Repairing the nets we always started on a double and finished on a double. They were diamond shaped mesh. The net was always called lint. When we'd repaired the nets they used to tan them. We had a big copper at the back, with a big fire under it. It was filled with water and there were boxes of cutch and they put this in the water and it made a rich reddy colour and then they'd dip all the nets in. That was what they called tanning. Then they'd hang all the nets out to dry, they were all cotton, no nylon that time of day. The Scotch nets were coarser, when they came from Shields they were a fine net and the nets off here were the fine nets, some had small mesh. The Scottish nets had big mesh. The fish size varied accordingly to where they were caught.

"The scene at the quay was so exciting. In Yarmouth you could walk across the river on drifters while the nets were spread all over the Denes. We used to go down the wharf when the boats came in and the men were so busy they'd no time to speak to us. All in their oily frocks, the sellers all shouting out, herring everywhere, we were slipping and sliding about. We used to have a carrier who brought all the spoiled nets home. His name was 'Mute' Goffin. He used to say, 'Your husband is in, put the beef pudding on.'

"We used to work from 1st September until the end of January then that fell off. If they went on the spring fishing we worked from May until the end of June again after that there was space until you came to the home fishing again. If we had bad weather then we had to work all the time, sometimes until nine at night. We had six girls in our room and had eight lamps with just one wick. However on earth we saw, I don't know, we had to clean and clean the lamps. It was tiring work as we were always standing except when you had a big hole in a net and had to shoot a piece of lint in then we put it over an oil drum and sat down until we worked it in. I've still got my needles and cotton. I believe there were about forty or more in Winterton and some women did the beating at home. When I finished I was earning 12/6 week. My husband was the driver in *Queen of the Fleet* (engineer) and his money was 17/6 a week. After the fishing the nets were repaired and thoroughly dried and stacked in the store. They were

Dennis George pictured at Winterton tanning the nets.
PHOTO: Mr. G. Wacey

A busy scene at the quay in Lowestoft with Yarmouth and Lowestoft drifters side by side.
PHOTO: Mr. P. Trett

carefully watched in case they sweated. They were all turned over and thoroughly dried. The ransackers did all the rope work putting on the 'norsels' which were the thin ropes attaching the net to the rope.

".....The first year he went to sea after I was married he brought in £5 at Christmas and that was for the home fishing. That was all he brought home. In the winter time I went net mending and he did the housework, else we had to manage on 12/6 a week....

"We used to get a lot of gales in the 1920s, it was definitely worse weather then. We really used to be worried when the boats weren't in. I've been worried about my own husband and we've sat and waited and waited and keep asking if the boat was in.

"When they went to Shields or Scotland we used to go up on the cliff and wave and they'd go 'toot toot' and some would have a tear or two. But it was a different tale when they came back. We'd have a letter from them to say they'd be back on a certain day and we would go and watch them steam by. We'd all get excited and walk off home to get ready for them.

"Winterton was a proper fishing village and we were all like one big family. In those days people were not unkind but strict. We were brought up to respect the parson, the schoolmaster, the policemen, all our elders in fact. If anyone was ill there would be twenty or so people go to see the house if they needed help. There was always someone to lend a helping hand and it's a pity that sort of thing has died out. They shared everything. If there was one or two fishermen come home Christmas time and they hadn't any money, the ones who had a little money had a party for the children and all the other people's children would go. Jim went to sea for fourteen years after I was married and I only had one voyage (good money) all the whole time he went. That year he brought £90 home. I put that in the bank and that's still there. I've never had any money but it's in the bank and I wouldn't touch it. I reckon that will come out soon! We had to manage. We had one night for ironing, clean all the knives and forks and spoons. On a Saturday I had to clean the toilet, clean and darn our stockings or they'd be holes from top to bottom.

".....Fishermen all helped in the house when they were at home, they had to, look how many children they had. One or two used to go up the pub and didn't bother about their wives but the majority of them were helpful. That's those in my generation, the older generation were different, they had a much rougher life in the sailing boats and first steam drifters.

"I knew all those old men, we used to have some laughs with them, well, we were afraid of them. They had such a rough living, they had to go to sea in those old sailing boats and the weather was a lot worse than it is now. They used to come home rough and ready, they were men for discipline.

We used to go down on the beach when the old men used to watch in the lookout and we used to go in the shed but if we saw any fisherman coming we used to run or we'd get a ding over the ear. We dare not go there when the men were on watch or anything like that. I can remember when my father was on watch in what they called the north shed. I used to go over there with a candle in a jam jar and some coffee in a can with a lid on. When they went away to the westward, about a quarter of the fleet went. It was rather risky, from the money point of view, some would risk it, others wouldn't. 'Wee' Green used to go and so did Grandfather Wodehouse. He only had one boat go once or twice and my husband, he never did go...."

I asked Mrs. Larner what the men did in between voyages. She told me, "They got 10/- poor money and they had to go down to Stokesby Pit and dig out stones for the roads. They seldom worked on the farms as the farmers had their regular men. They nearly all had their own allotments though and they worked on them but they had to depend on the women. Some used to go to Grimsby, trawling, in January and stay until about March. There was a lot of fishing off the beach and there must have been fifty little boats when I was young. Every house in the village had what you called a 'speat' which was a long narrow piece of steel. All the year round you'd see fish hanging out drying on them. That's what fishermen principally lived on, fish. In winter time they used to take lines on the beach and they used to catch no end of cod. My father used to have a little barrel and he used to salt all these cod down. On Sunday my mother used to put them in water and soak all the salt out and on Mondays, washdays, they used to boil it and make the white sauce. They used to call it melted butter, they all used to have that on Mondays. The rest of the week they had to get what they could.....

"Really and truly, old people in those days always found up a good meal somehow or other. At sea they ate mostly fish, some of those men ate ten or a dozen herring at a time. Fishermen used to eat huge meals, my husband used to. Fishermen never used a knife and fork either. They used to 'snotch' the herring all down and used to take them between the snotches and put them in the mouth, all fried nice and crisp. We used to pickle them with cloves and vinegar. They salted all cod for the winter and red herrings. They also had potatoes from their allotments and half of pork and pea soup.

"The women used to get 2/6 for knitting a jersey. They used to put fancy cables, they were right fine navy wool and they used to have sixteen needles. They didn't have a special pattern and used to make their own patterns up. They'd sit on Tower Hill in the sun, there used to be hot summers then, Mandy, Giffer, old Mrs. Tinman and all them.

"One time that was blowing a gale of wind, most of the boats were in and they said Jim's boat wasn't in. I know I was worried, I stayed up to 11pm and then a knock came on the door and a man brought Jim in. He'd run ashore inside the piers, the boat sank, he was then engineer. He never went to sea after that. That was in the *Achievable*. I've got a painting of her that my husband painted.

"When he was in the '28' (one of Smith Dock boats), she was an iron boat. She shipped a sea off here, they were all battened down but they went down and they all thought they were going to be drowned. Poor old Mush was on her and he got on his knees and prayed. They thought she wouldn't come up again but she did come up. He lost one of his pals who was with Harry Powles. He fell overboard. They got him but he was dead. I think it was *Harry & Leonard*, one of Eastick's boats.

"Fishermen use to watch the barometer, the tides and the weather. They were wise men. They used to watch for all the signs. There were not so many caught in small boats by the weather like there is now. They used to come ashore in bad weather. Nowadays they get caught.... We were worried when the men were away. Not when they were in Scotland because that was in the summer time. It was the autumn that was the worrying time. Fishermen took it all in their stride good or bad weather. The autumn and winter was the worst and we did used to worry in the bad weather....."

Chapter Four:
A HEAVY SEA

In 1900/1901 Dennis George went into steam drifters and his memories tell us much about the early days of steam. Many drifters still relied on sail though, the engines were low powered, little better than auxiliary. Thin woodbine funnels, no wheelhouse, or the wheelhouse aft of the funnel, but with improved accommodation and less dependence on the weather and tides.

Dennis recalled, "....Old Dick Sutton, he was a little short man. He had one of the first steam drifters in Yarmouth named the *Fancy*. I went in her after she was about six months old and I brought home over £70 with the home fishing which was extra good. All in gold sovereigns too, lot of money in those days.

"In 1905 we were 56 miles SE of Lerwick and most of the boats were fishing E by S 80 miles from Lerwick. We were fortunate that night we got 50 or 60 cran and it came on a heavy gale of wind. The Smith Dock Company was one of the first fishing companies at Yarmouth and their ships were named from *One* to *Thirty Four*. The *Twenty One* belonging to them went down, a man named King from the north end of our villlage was skipper and he was lost. The rivets fell out but the crew all jumped into another drifter and were saved."

Dennis George like all fishermen made light of such a rescue from a ship sinking in heavy seas. "We set the mainsail, to'foresail and mizzen. In those little steam drifters we had a mainsail, to'foresail and mizzen. We'd pulled up half a fleet of nets on the starboard side and all of a sudden this gale of wind caught us and took our mainsail away just like that, a good job it did else she'd have gone right over the nets being all on that side. The mainsail blew away but the to'foresail didn't so we put her before the wind. When we got to Lerwick we had to let go our big anchor and wear down to get our sample in and we were the only ones which landed herring that day, all them out E and S never got any fish and had to dodge the weather."

Dennis mentioned a mishap which sometimes occurred on these early steam drifters. "There was a Lowestoft boat, Hodges was the skipper of her but I forget the name of the boat. One of the crew came to him in the wheelhouse and told him that the propellor was lost. 'Oh,' he said, 'we're still moving though?' The chap said, 'I mean the one on the afterdeck....'

The propellor would sometimes drop off or get damaged so a spare one was carried on the afterdeck. Dennis continues, "…..It was fixed with a bolt that used to come through the deck with an iron rod and a screw. What happened was when she fell off, the sea then broke that bolt and as she forged ahead the propellor went right over the quarter and they lost it, all through that gale of wind and the big seas."

Another gale that Dennis encountered took place during the home fishing season. "…..When I was in my own boat in 1911, the gale of wind came down on the last day of September. We got 10 or 12 spoilt nets and about 80 or 90 cran of herring. All these spoilt nets lay about the deck and when that wind caught us it took a lifebuoy off the top of the wheelhouse with 15 fathom of line fast to it. We were 80 miles from Palling and it was 12 o'clock on the Saturday when our lifebuoy blew away and it was 8 o'clock the Sunday morning when it was picked up at Palling. It had blown 80 miles in that time!

So many boats were lost without trace. This Lowestoft sailing lugger is coming up well reefed, but sometimes in a heavy gale the ballast used to shift from one side of the boat to the other with disastrous consequences.

"The *Gal Nancy* was speaking to us when that wind came, we never saw her again and the *Montrose* was sunk not far from us and five got lost out of her, one a Winterton man, while another drifter saved the other three or four men. Then our traveller broke off the mizzen boom, we had another new one down the fo'castle. It was some time before I could make them understand what to do as the mate was only twenty-one and there was only one who'd been in a gale of wind along of me, the rest were only youngsters. I say put a line around my waist and I'll go out and reeve this other one, but it was sometime before we got it reeved and got it right. I gave a Yarmouth chap the wheel and said whatever you do keep the wind on the starboard bow and don't let it come on your port bow and he did. In that weather the wheel took charge of him. On the wheel we had a brass spoke to help it round and it struck him in the arm and broke off. He stuck it all the time but his arm came up like anything. We dressed it when we got him down below." The men of the *Montrose* were saved by Yarmouth drifter *Piscatorial* which was later sunk in 1914-18 war.

"We got stowed up, two tons of coal we had on deck and we had to throw it overboard. We dare not put it down below. We put the nets down below and got battened down. The next day we turned out and salted the herring we'd battened down. That was on the Sunday and I say to our chaps me and the driver have been about for several hours, we will turn in till 12 o'clock and then we will start to go home. The mate and the another chap had the watch and we turned in and got up half past four and so to Yarmouth.

"In the following year 1912 (there was a flood that year) on August 12th we got caught in another gale of wind about 40 miles from the North Haisboro, that didn't last quite so long but we broke the rigging that time."

Dennis George like many drifter skippers took part in rescues and salvage operations. Here is the description of an incident with a full rigged sailing ship in the Bristol channel. ".....When I was in the *Superb* in 1905 after mackerel we used to go miles from Milford Haven and off Kinsale and we had eight or nine thousand mackerel on and we came out of Kinsale to come to Milford. When we'd come a hundred miles there came a gale of wind we had to dodge through. At four o'clock of the morning we see a ship, a full rigged ship with no sails up, a four masted ship a running before the wind. I say to our chaps, 'He ain't going to clear the Smalls, he'll run ashore.' That was about thirty miles off Milford and she was on her broadside. As we bucked up a bit to go to Milford, they hailed us to tow them in. The tug that was towing them from Hamburg to Port Talbot got the hawser round her propellor and was lost with all hands.

"There were seventeen Italians aboard that ship and only one could speak English. They slacked their tow rope away but we couldn't get hold of it and they blowed right in towards Caldy lighthouse. We kept along with her and threw a rope with our lifebuoy aboard and they all held hands ready to jump all at one time but they didn't do it and they wouldn't let us haul them back and forward with our life line. They were within a mile of the shore blowing a gale of wind and I went forward and held our little anchor up signing to them to let go their two anchors and they see that and let go both anchors and they held. At 12 o'clock it fined away and I went aboard her and the captain asked me if I could fetch him a tug.

"We steamed 32-33 miles up to Swansea and when we were off the Mumbles we see a tug and spoke to him and told him what had happened. He steamed back inside the Harwick Sands and we steamed 36 miles back again. When we got to them the tug hadn't got a tow rope so he had to go back again! Well, I went aboard and I couldn't get 40 lbs of steam up on the old boiler to heave the anchors up, they only had like old sand for coal, so I say to him will you let me try to get steam up. The bloke who spoke English said yes, so I got it up to 90 or 100 pounds and we put the chain on the capstan on the after deck and hauled the anchors up and they were both foul. They had an eye bolt on the forward deck where they could make things fast when the anchors were up. So we put the stoppers on and put through this eyebolt and slacked the chain up so they could get clear but the eyebolt pulled out of the deck and both anchors went down. We pulled them up again but still they were foul. I put a five inch rope through the stock of the anchor but that broke and let the anchors down again, they both fell clear so we put the chain on both anchors on the capstan and pulled them up together. I kept the chain on the capstan and made fast. I say to the chaps, they'll have to pull the capstan off now if there's too much weight. We towed her twelve miles until two tugs came and he insisted they tow him. We steered her because she was swinging about on the broadside and we helped to tow her to Port Talbot. I gave them two or three trunks of mackerel and they cooked them, heads, guts and everything. They just didn't know how to cook them...."

Dennis George gives just a factual account but reading between the lines we realise it was a stirring example of the sailor's tradition of standing by and assisting others in distress. In such weather on a dangerous coast, it was undoubtedly hazardous for a small fishing vessel to board and then attempt to tow such a large vessel to safety. The *Superb's* part in saving the ship was recognised at the subsequent Admiralty Court hearing in London.

A Lowestoft vessel, the Race LT35, *is shown dramatically caught in a heavy sea in the 1930s. On the vessel can be seen a radio mast. These were mostly receiving sets so the weather forecast could be heard.*
PHOTO: *Lowestoft Maritime Museum*

Another time a fair sized sailing vessel came ashore and was wrecked. Among her cargo were many jars of sarsparella, a well known drink years ago. The people of Winterton, like all east coast dwellers, regarded wrecks as acts of God, not to be wasted but salvaged for personal use as much as possible before the coastguards took charge. In those days toilet facilities were pretty basic and the George's had two earth closets side by side so they took the jars of sarsparella and put them in the unused closet and buried them, using up an obviously sensible disguise! They were never found!

Winterton had several lifeboats with the beach companies along with the lookouts. They were very active in the sailing ship days as it was a dangerous coast for ships driven towards a leeshore. Many of the Georges served in the Winterton boats. Apart from assisting ships, salvaging and life saving, one of the main activities was 'swiping'. I was puzzled until Dennis explained, "Swiping for anchors, we had big grapnels and brought them up and ashore." The sailing ships beset by gales and in danger of driving ashore or on the sandbanks threw out their anchors to hold them fast until the wind changed. For one reason or another many anchors were lost and it was a very profitable activity, recovering and disposing of them.

Chapter Five:
THE NEWLYN RIOTS

In 1896 Dennis George witnessed perhaps one of the most notorious fishing riots in this country's history, which ocurred in Newlyn, Cornwall. Dennis explained, "....The locals strongly, and rightly objected to fishermen from other parts of the country fishing on Saturdays and Sundays. It leads to overfishing and low prices and feelings ran high.....

"I was in the *Nellie Jane* along with Jimmy Smith and we lay in Penzance Dock and there was about twenty of us in there and we went out on a Monday morning in a flat calm. We set the two big topsails and the spinnaker and we only got seven or eight hundred yards from the pier when three chaps came along and boarded us and seemed friendly enough. We hadn't got a mile from the pier and up came twenty or thirty in their little sailing boats, 'nickers' we called them, and they rushed aboard as hard as they could, knocked the skipper away from the tiller and say to him, 'Get back!' I was net stower and my pal, his name was Smith, the owner's brother's son, said if we'd known they were going to jump aboard like that we'd a had the hatchet ready to chop their fingers off and we would have done that too!

"They said to our skipper that they were going to take us to Newlyn where they had a chain across the river. Then he told them, 'Look here, we'll go back to Penzance dock if you let me have charge of the vessel.' They agreed and gave the tiller up, as long as we went back. We laid there in Penzance dock, about twenty of us, Monday, Tuesday and Wednesday. On Wednesday afternoon a man known as 'Suet', skipper and owner of the *Mizpah*, put his little boat out and came against the pier to take so many blocks of ice for the mackerel that he had to take to Plymouth. They used to bring ice down in 1 cwt blocks, there was a ship loaded with ice that came from Russia to Penzance. Suet was alongside the pier with two men and they'd only got two blocks aboard when the Newlynders came rushing along the promenade on to the pier. There was a boy in the *Rovers* boat going to have a scull about the dock but they got in and used their hands like oars only they couldn't get alongside Suet's boat. He saw them coming and let go the spinnaker sheet, went to leeward and a little draught of wind came and got away and he stood at the tiller and slapped his backside at them. The skipper of our boat said if they try to get alongside us we would get the anchor and throw it into them and sink her but they didn't get alongside. The Lowestoft skippers were all on the quay and said if they hurt that boy we'll have to do something, but they didn't harm him.

"Then we came ashore and the Penzance people were all against the Newlynders and they sent for the excisemen to come from the Scilly Isles to have a meeting about the Saturday night and Sunday night fishing. When they got to Penzance there was a riot and some nearly got killed."

"When the Newlynders came on the quay a butcher happened to come along with beef for the Lowestoft luggers and they took his beef basket from him and threw it over the pier. There were two landings at Penzance, he got on the top landing with his big knife and said, 'If you come up here I'll rip you up like beef off the bullocks rump....' It seems that they thought he was something of a traitor as Dennis explained, "The butcher shouted if lots of you paid me what you owe I could retire tomorrow...." and he wouldn't have needed to sell beef to the Lowestoft men.

Dennis continued, "We lay there and the Penzance people got to work with pokers and anything, you never saw such a sight in your life. Some of them nearly got killed on that pier. Wednesday night me and my friend went for a walk on the promenade and I said, 'Shall we take a walk as far as Newlyn harbour?' That was about a mile. So before we went we took off our thick silk wrappers and tied a stone in the end to have something to protect ourselves with. When we got to Newlyn there wasn't a soul about anywhere, they'd all turned in.

"There were seven or eight Lowestoft boats in the harbour and there was a chain across the river, they were all locked in. We spoke to one of the soldiers on guard, and he told us on the the first day of the riot that one chap ran afoul of him so he cut his ear off! Then he told him to pick the ear up or he'd cut the other ear off too! That was a very rough do but there was no more trouble.

"We set out on Thursday morning in company with our other three boats in strong wind and we shot the nets and the next morning we were fortunate to get some three thousand mackerel. We had 1000 each and the skipper said we better put them all together, take them in and get a few pounds so they put all the catch into our vessel and we sailed to Plymouth. When we got to Plymouth there were thirty ships, most part luggers, that had left Newlyn. After we'd landed, it was a Friday, the government bloke said, 'I want all you sails to follow me to Penzance Bay just after tea in the evening time.' The Newlynders were all there on the pier. The government boat steamed about half a mile off the pier, us lot following him up, and he fired two blank shots. Well, they rushed about and were out of sight in no time. The government ship put a boat out and they unlocked the chain across the harbour and that was the end of The Newlyn Riot of 1896.

"When we went to Scotland or Ireland that was the rule, no fishing Saturdays or Sundays, and we had to abide by the rule. It was commonsense and probably if that had been the rule in Yarmouth we wouldn't have had to catch so many herring and could have got a better price. There were many times when Yarmouth fish wharf was full of herring and still we went out on the Saturday and Sunday nights and came in right full. The fish salesmen were to fault. They didn't think that half the number of herring would make the same price. Later on we came in with a hundred cran and could only sell thirty cran, so we only went to sea every other day after that.

"1930 was the last year for making a profit, I only had one boat then. Local people lost every penny. There were thirty boat-owners in Winterton. Some had one boat, some had two, some had four. I sold my boat in 1936 for £750. Billy Balls bought her. She was lost during the war and he got a good price for her. 'Starchy' John George, who was my relation, and Fred Goffin along with others who lost boats during the war were the only ones who had anything worth having. In Yarmouth and Lowestoft no end of owners lost every penny who in good times could have retired on £15,000 or £20,000. They lost everything...."

An early drifter at Yarmouth harbour, with a woodbine funnel and a full set of sails.
PHOTO: Mr. P. Trett

Chapter Six:
'STARCHY' GEORGE

Dennis's relation, John 'Starchy' George, was also involved in the fishing industry. He went to sea all his life and eventually became owner and master of his own boats. He also had some interesting memories of fishing in days gone by. He recalled, "....The sailing boats were always called luggers. They were sailing drifters and then there were the smacks but they went trawling and the luggers were herring catchers. They were rigged practically the same, the smacks had a boom to the mainsail while the luggers hadn't. They had what they called a 'horse', the sheets used to slide back and forth on an iron beam. The smacks couldn't lower their masts but when the herring drifters shot their nets they used to drop the foremast down to an angle of 45°. That's the difference between a smack and a lugger.

"The majority of sailing boats went on to about 1903, sailing and steam boats together. I remember a sailing boat going down to Lerwick in the Shetland Isles about 1907. The skipper's name was Ted Stolladay. There were sailing boats in 1909 when I first went master, just one or two but steam came in more or less at the beginning of the century. The first steam boat had no wheelhouse, no galleys, just a small funnel stuck up in the middle, a small boiler and an engine. They'd bunker about eight ton of coal, in fact, the *Lottie* when I was in her had wooden side bunkers and just a little 15 hp engine. They were more auxiliaries than anything. We used to do a lot of sailing. We had a mainsail, a to'foresail and a mizzen. With a fair wind you'd always have them set, they'd give you 7-8 knots in fine weather.

"The *Lottie* was about 70 feet long and about 17 feet beam and drew about 9 feet of water. There were ten in a crew with a cabin and a fo'castle, four in the cabin and four forward, the cook had to sleep with one of the men and one slept on the locker, eight bunks for ten people! A lot of boats used fo'castles until 1914 and some after that. Most of the steam drifters hadn't accomodation for all the crew in the cabin. Some had a berth in the wheelhouse for the skipper. The *F and GG* had, and *Rose and Gladys* and the *Harry & Leonard* they did, but they were too hot being so near the boiler.

"From then on it was all steamboats. I believe Bloomfields had the last steamboats built in 1930, *Vim* and the *Lux*. The early steam drifters cost

The Consolation *was the first steam drifter in Lowestoft.*
PHOTO: *Lowestoft Maritime Museum*

about £1900. At Lowestoft they'd be about £100 - £200 cheaper. Iron boats were dearer. When Bloomfields built the first dual purpose motor boats, *Ocean Starlight* and *Ocean Sunlight,* they cost about £45,000 and that was small as even then the prices went up to about £70,000 and that didn't include the gear!

"The early nets were the English rough nets and then they got the Scotch nets, a finger net, cotton, although now they're different and are nylon. I used to pay 75/- then but by 1956 the cost of the net was about £17.50. We had about three hundred nets for a drifter and she carried about eighty to a hundred. Each net was about 34 yards. We'd shoot about ninety, that was about one mile of nets lying ahead of the boat. Previous to that, in the sailing boats, they used to shoot a hundred and sixty to hundred and seventy nets. They were the old rough nets, different from the

Scotch nets, they were shorter and had no rope on the bottom; the Scotch nets had rope on the top and bottom and corks on the top. The rough nets had small wooden barrels called bowls before the pellets came in.

"The nets were the biggest worry especially after they got Scotch nets and steam drifters. They used to go to sea in worst weather and shoot the nets and if they destroyed twenty or thirty during the night they all had to be taken ashore, be dried, taken up to the beating chamber, be mended and go through the ransackers' hands again and that was a big item. The old rough nets would stand the weather better.

"The steam boats went to sea in worse weather than the sailing boats. My father used to say to my brother and me, it won't be too bad, I should go. When the herring season was in its prime everyone was trying to be top boat, trying to make the most money. Some of the owners were just as bad as the skippers, you know, were almost driving them to sea, ordering them to go, 'It won't be so bad,' and away we'd go. I don't think many people worried about the boats. You never thought about a drifter getting lost. I hardly ever heard of a drifter getting lost, not through stress of weather, it was the nets you worried about, if you lost or spoilt many.

"The beatsters were women who looked after the nets, the lint as we called it, and the men, the ransackers, they looked after the ropes, that was at the store. The nets all had to come ashore, be dried and repaired.

"When I first went master, I bought coal at the Tyne for 10/- per ton. In later years best Yorkshire 'hords' were 29/- per ton. In the early days when I first went to Scotland in the *Lottie* with my father, I was second engineer at the time, we had 49 tons of coal out and back for eleven weeks. That's about $4^{1}/_{2}$ tons a week. As I said, we used to do a lot of sailing as well as steaming. Afterwards they didn't use a sail and later didn't have a mainsail, they have to have a mizzen to keep her head to wind. Later years they'd burn 15 or 16 tons a week easily and we used to say a cran of herring paid for a ton of coal.

"The owner took the cost of food, oil, coal, ice and salt etc, out of the gross earnings, after which it was divided in nine/seven shares, nine shares for the owner, seven for the crew (later it was 50/50). The master had $1^{3}/_{4}$, the mate $1^{1}/_{4}$, engineer 1, the houseman 1, younkers $^{7}/_{8}$, down to the cook $^{1}/_{2}$. A crew of ten or eleven, all paid at the end of the voyage. If it was a bad voyage they ended up in debt.

They used to have a joint of meat a day, 5-6lbs weight. When the cook got used to it he knew within a little what was needed (boys always started as cook and in the rope room, both very important jobs and they seemed to learn quickly).

"If you couldn't sell them at all you'd throw the herring overboard. We had 120 cran and sold 40 at 2/6 a cran and the rest we had to dump. If you averaged £1 a cran you were doing fairly well. Prices varied accordingly to how many herring there were and the demand. The biggest shot I ever haulked was 236 cran. They get more now because the motor boats have more storage.

"Since I've been owner and master, I've never been mackerel fishing but I used to go when I was manning the boat. We used to go to the westward and right round to Limerick on the west coast of Ireland. We'd sold them at a little place called Finnet and we used to come round to Kinsale, Castle Bay, then back to Newlyn, Penzance, Plymouth and back home.

"Yes, there were big Atlantic rollers, very impressive in a gale, but the North Sea was worse, it was a sloppy sort of sea. In a gale of wind and heavy seas we had to 'dodge' in the North Sea. Dodging, you set the mizzen about half way and got at half speed or quarterspeed to keep them head to wind and you could ride some funny seas then. Yes, the drifters were good sea boats. As I said, I hardly remember hearing of a drifter being lost through stress of weather, very few indeed.

"In the early drifters we only had a compass and a weather glass and they all had a leadline. The old skippers years ago knew exactly where they were by the lead. It was hollow in the bottom and they put a bit of fat in. They'd hit the bottom once and whatever was in the lead would tell them where they were. There are various places in the North Sea where you get soil you don't get in other parts. How they kept it all in their heads I don't know as most of them couldn't read or write. When the steam boats got about old Harry Myhill, master of a trawler, said, 'It's no good taking a east in the North Sea now, it's all cindermuck' (which was ashes and clinker thrown overboard).

"In the old days, before the modern aids to fishing, we used to look for the colour of the water, watch for any sea birds, gulls or gannets. Sometimes the water would be a dark dirty colour and you knew there'd be no herring there. We'd go where the water was milky, as we said, greasy on the top, you had to watch the water. You never got herring in the sheer clear water.

"Boats in the early days they were all local owned. Some owned one or two boats, some had half shares and some had six or seven boats. A lot of wooden boats were built in Yarmouth and Lowestoft and several were built round the West Country. There were not a lot of iron boats built at Yarmouth and Lowestoft at first. A lot of iron boats were built up north. The Smiths Dock Co at North Shields and the Scotch boats were mostly built in Aberdeen and on the Clyde. Many wooden boats were built at

Yarmouth by Fellows and Beechings and at Lowestoft by Chambers, Richards and Reynolds. Between the wars they built the *Lydia Eva* (now the sole surviving steam drifter) at King's Lynn and one or two boats for Lowestoft. Several were built at Selby in Yorkshire. The engines were built by Crabtrees, Calvers and Burrells at Yarmouth, Richards at Lowestoft and Elliott and Garrood at Beccles.

"Ninety-nine out of a hundred capstans were built by Elliott and Garrood at Beccles, who also supplied propellers. Between the wars dozens of them stood awaiting demand outside the works at Beccles. The little boat on the stern seemed to store all sorts of things, cabbages, potatoes etc. They were seaworthy, they were looked at, especially after they had the Board of Trade survey of 1938, they were all tested. The boats had to be surveyed, the engines, the boilers, the surveyor looked at everything and if needed it had to be done, steering chains, propeller shaft, stern tube, gudgins on the rudder had to be replaced if worn. Every now and then a propellor would fall off too.

"In the cabin you had a bunk, just a hole in the side, bareboards. You had to find your mattress and blanket. Your shore going clothes were in a kit bag around your bunk. Your sea-going gear would be in the lockers under the bunk. Your oilies you sometimes hung down the ventilator. The only thing to wash in was a bucket on deck or in the engine room if the engineer would let you. You'd hardly wash in the cabin as the cook would get on to you as he had to keep the cabin clean. Fishermen going ashore wore a blue serge suit, a guernsey, close knit with fine wool on those little needles and very warm, a warm shirt, no collar and tie, and shoes. You had to look after them.

"When I first went to sea there was no lavatory, only a bucket which some people took in the wheelhouse. On the early drifters there was no wheelhouse, just a bucket on deck, and you managed as best you could.

"Fishermen were superstitious. You must never mention pigs, rabbits or parsons and they wouldn't have a parson come aboard. Boats would not start a voyage on a Friday 13th or any Friday.

"Scotchmen never went to sea on a Sunday but the Englishmen did. Cornishmen wouldn't go to sea on a Sunday, nor the Irish. When the English went to Scotland or Ireland however, they didn't go to sea on a Sunday either. Things are very different today with aids to navigation, fish finding and every modern convenience. When I first went to sea, a man was very much master of his ship and some were proper old tyrants. If things went all right, they were all right but if they didn't you knew it!"

Dennis George on an early drifter, the Young Archie *in the early 1900s. Dennis is the one standing in front of the funnel and notice the little lifeboat to his left, full of gear.*
PHOTO: Mr. G. Wacey

Chapter Seven:
FISHING FOLK & BOATS

During our conversations Dennis George recalled many fishing personalities, "....Old Dick Sutton was a little short man. He was a shrimper and his brother Harry was a shrimper. Old Dick bought the *Adviser* and then had up to four drifters. He had one of the first steam drifters, the *Fancy*. I went in her after she was about six months old and I brought home over £70 on the home fishing season which was extra good. We got over £2500. He retired in 1914 on £36,000. Well, after a year or two he had a motorboat built and she was a failure. Then he had two big steam boats built like the *Frenchmen*, which carried six or seven hundred cran of herring, and barrelled the herring at sea. The *Rose* and the *RRF* were big boats with drifter lines about 130 ft long. He had those two and he thought things out and had 'seine' nets instead of drift nets but that was a failure too. He lost £14,000 due to his experiments. Then he sold one of the boats, the *RRF*, to a Frenchman and the *Rose* to 'Crabby' Hudson of Filby. I went and valued the nets up for him for the bank and a Kessingland man valued them up for old Dick Sutton. 'Crabby' did fairly well in her. Old Dick Sutton lost £14,000 in a very short time, if he'd kept on he'd lost the lot.

"The two big Sutton drifters were much longer than the usual run of drifter. Another unusual drifter was the *Lydia Eva* (YH 89), one of the Eastick fleet built as a drifter/trawler at King's Lynn in 1928. She was bigger than the traditional drifter with a big bluff bow. These three were distinctive but not traditional steam drifters. A true Yarmouth drifter was the *Wydale* (YH105) a wooden ship built by the famous Lowestoft ship builders Chambers, a good ship and a lucky ship, fishing out of Yarmouth until the very end of herring fishing. If ever a herring drifter deserved to be preserved it was the *Wydale*, a typical wooden steam drifter. With skipper A. Brown she won the Prunier Trophy in 1950 with $250^{1}/_{2}$ crans. Actually she had 300 cran in her nets but handed some nets containing 60 crans to the *Harry Eastick*. The *Wydale* was a worthy representative and a working last survivor of the great East Anglian herring fleet. In 1961 she was sold to Holland for breaking up, a sad and sorry day. Another drifter I liked was the *Wilson Line*, another Eastick boat often fishing out of Yarmouth. She was an iron boat, later converted to diesel."

Ernie Brown was one of the famous Caister family. In the 1914-18 war he was an engineer in the Navy but he had many memories of his fishing

Steam drifters at the fish wharf in Yarmouth in 1955. The vessel the Wilson Line *was sold in Scotland but continued to operate out of Lowestoft and Yarmouth.*
PHOTO: Mr. P. Trett

days. He recalled, "I went in a Yarmouth boat, then in the *Three Kings* out at Lowestoft, then in the *Adelver* one of Baker's boats out of Yarmouth. She was a government boat and they bought her, she was a steel ship. I was in her for about two years then I went into one of Green's boats (the famous Wee Green of Winterton). He was eighteen stone and he had a little old Austin seven car. I went fireman for one year, then driver and stayed along of him until the finish. I ended up in the *Wilson Line* until they put a diesel in her then I finished. When I left the sea I went with the Port and Haven Commissioners on their dredger and had two years in her.

"Steam drifter engines were compound and triple expansion. As they got on they were a better ship. That *Wilson Line*, she had a lovely engine. I remember the *Wydale*, she had an Elliot & Garood engine from Beccles, a monkey triple they called it, one cylinder on top of the other, they were a compound but a triple engine. They only had two legs instead of three. I was in two or three of them, *Ocean Swell*, one of Green's boats was one of them. He was skipper and boat owner then he went with Bloomfields as ship's husband. I was in the *W.G.P* and next year he came out of Bloomfield's. Took his boats out and went on his own then. Lennie Dawkins followed him as ship's husband for them. Wee Green asked me then if I'd go driver in the *Ocean Roamer*.

"The coaling and keeping a head of steam, that was a job, especially on the long runs when I was in the *Rose Hilda*, that was another of his boats. 240-250 miles, that was hard work and she could carry over twenty tons. She had a big cross bunker and we had to clean fires every four or five hours. One boiler and two burners; you ran on one while you cleaned the other. I liked steel ships better than the wooden ones. They were more steady in the water, they wouldn't roll about like the wooden boats. Steel boats were the best – you didn't have to worry about any leaks in them.

"Those steam engines were reliable all the time, I was many years in them and I never did have one breakdown. Two years I went fishing off Dunmore, we landed the fish at Milford, winter fishing after the Yarmouth fishing. We fished round the Fastnet, 50 or 60 miles, that was hard work. I had to help hauling if there was a big shot. I chose my place so I could work the engine if they needed to go ahead. As a driver I got $1^{1/4}$ share. The skipper got $1^{3/4}$. the mate $1^{1/4}$, engineer $1^{1/4}$, 9 shares for the owner, 7 shares for the crew which, later became 50/50. There were ten in a crew. Later when the boxing came in there was the hausman's 1 share, down to the cook $^{1/2}$ share. The cook coiled the ropes and sorted them out of the rope room. That was a job where you could get muddled up! I've been in a muddle when I went, and they keep coming, all rope and no room. You had to coil them neat or you couldn't get them in.

"We had a fireman and we used to have watches. When I first went we only had short journeys about forty miles. The fireman used to take the ship to sea and bring her home. On the longer trips we used to take spells for three or four hours. The propeller used to race when we were running before a swell especially if we had a big shot of herring and the bows were down. I was in the boat *Adelver* and went in to Yymuiden in Holland with a big shot, about 290 cran, and when we got in we looked over and could see one blade of the propeller, her head was right down.

"The worst sea I remember was when we were seine netting, this was years ago, fishing the Minch on the Scotch voyage. We were off the Dogger Bank making for Shields when we met a heavy gale of wind. She went into a sea and I never thought she'd come up and she sprang a leak and leaked so much we had a hard job to keep the water down. We had to go in dock at Shields and all the foredeck had lifted. She'd knocked all the 'oakum' (the packing between the deck planks) out of her from her wheelhouse forward. It all hung like rags. I've never seen anything like it, we were five days in dock.

"I never had a breakdown. I was lucky although there were very seldom engine breakdowns in a drifter. They were very reliable. I was towed in once. I was only fireman at the time on Dennis George's ship the *Norford Suffling*. She was brand new, all new ropes and everything and we were towing the ropes astern. Larner was the skipper and we got the ropes round the propellor wound up tight and we couldn't free them. We got towed into Shields by a Grimsby trawler and that's the only time I was towed in." This vessel is still afloat. It was converted in 1983 to a sail training ship in Denmark and renamed the *Grethe Witte*.

"Dennis George managed the *Norford Suffling*. We had to watch the engines at slow speed when we were dodging, especially with the Elliott and Garrood engines. They used to stop, only having two legs. They used to centre and we had to pinch them over. The ordinary drifter engine had triple and they went slow but never stopped, they kept jogging over. We had a telegraph and a tube and some rum old language sometimes came down that, if we were too slow or too fast!

"I used to look after the gas lighting. It was all through the ship. We had a big generator, a gas generator, two boxes. You used one and when that was done went on to the other carbide. We still used the oil lamps for the steaming lights. I liked the home fishing best but I didn't mind the Scotch fishing. It made a change...."

Elliott and Garrood of Beccles, mentioned by Ernie Brown, were quite a big name in fishing. Dennis George could remember how the partnership was forged, "......Garrood was an old friend of mine and I borrowed money off Elliott when I first started. Garrood was a blacksmith and Elliott was an engineer before they joined together. Old Elliott was walking past a shop with a Singer sewing machine in the window and thought he could make a capstan like that and he made a capstan that brought the rope in. It had a little engine on the top, a little steam engine and he put a boiler down in the cabin in the sailing boats and it was led up to the capstan and we heaved the rope in with that and we used to have a little governor go on and if that broke we used to put a rope one on. All you got to do was turn a wheel on the top and off she go round, four wood blocks and four iron ones where the rope go round. That was stood near the hatchway to the rope room and the rope would go there for the boy to coil. It was marvellous and he sold thousands (the famous Elliott and Garrood capstan).

"The *Nellie Jane* that we were in had a big cabin, she was a little boat they'd bought from the westward, other side of Falmouth. She'd been a little steam drifter, a steamboat. Where her propeller had been was chocked up and 'Mouse' Catchpole came aboard and wanted to know if Jimmy Smith, the owner, would let him put a steam engine in her, he only wanted so many feet of the cabin but he wouldn't let him do it. So then he got this other one built and named her the *Consolation* (LT718), a wooden boat about 69 feet long and this was the first steamer out of Lowestoft. She had a tiller aft, no wheel then, no wheelhouse, nothing, yet the first one I went in, the *Fancy*, had one in 1900s. Mouse Catchpole did extraordinarily well in her and old Elliot said run her for the home fishing and if you're satisfied with her, I'll give you the chance to buy her. Well, he bought her and did extraordinarily well."

Apart from his fishing memories, Dennis George remembers much of Winterton's history, told by his family elders. In the time of Nelson, the navy was always short of men and the press gang was very active. One morning the look-outs at Winterton spotted a naval vessel drawing near the beach. All the younger men guessed what that might mean and wasted no time in rushing inland and hiding, that is all, except one of Dennis's forebears. This one young man said, "They won't take me," and refused to flee. A boat from the ship landed on the beach and found only an old man, women, children and that one particular young man, who seemed strong and active. "We want you," they said. He replied, "I want to come with you," and they were more than pleased to get just one, and a volunteer at that. Unable to find any other suitable men the sailors returned to the ship. When alongside he climbed straight up the side, looked around and then promptly raced up the rigging to the crow's nest where he began to cry, "Caw Caw," very loudly. At first this was regarded as a bit of a joke but it created such an uproar that it roused the captain. He nearly went mad and called the boats crew everything in his extensive naval vocabulary. The captain shouted "You have brought me a raving lunatic. Get him down and take him home!" and they did. Later when the ship sailed on and the men returned, the young man said, "I told you they wouldn't take me...!"

Two vessels being towed to sea.

Chapter Eight:
STEAMING HOME

The most difficult harbour Dennis George ever approached was Wick in Scotland. He recalled, ".... If you had a strong east wind blowing straight in you had to turn quick on your port helm to get round. There were always men on the pier to look out and throw you a rope to check you round, it was a very bad place to get in. North Shields was also a very bad place when it was blowing east and so was Yarmouth when it was blowing a hard east wind. Lerwick, that was an open place, when you got in it was a mile and a half wide, plenty of room and if you wanted to put your sample ashore and it was blowing hard, the wind coming the blowsy side to the quay, you had to let down the anchor and chain to slow you down down so you could put your sample ashore for the salesmen."

It was a difficult job sailing into Yarmouth harbour too. "....If the wind was right we used to sail up and let go the anchor before we got alongside the quay. If there was an eastwind blowing hard with a strong tide we used to put spoilt nets overboard and a couple of bags to slow us down so you didn't run on the sea. If you ran on the sea the same speed as the sea was going, you couldn't steer then. We put the gear out so went slow into the harbour. If the wind was east you could sail right up the river but if it was north you had to beat and then let go your anchor and wait for the tug. The tug towed us in and towed us out about four at a time...."

Today there are many aids to fishermen but in the early days, apart from a compass and the barometer, they relied on experience, local knowledge, and especially, the leadline. As one retired boat-owner and former skipper said, "I don't know how they did it, but they did. They knew their way about." Others simply said they 'smelt the bottom.'

"When you're on this coast you use the leadline, you put fat on the lead, bring the soil up and you know where you are within a few hundred yards. You have your chart, you have your lead and you know where you are. When you went 25 miles NE of N of Smith's knoll, you put your lead over and brought the soil up. It was all like little 'Bishee Barnabies', you put them on your teeth and cracked them, if they were hollow you knew you were 23 or 24 miles NE of N of the North Haisboro."

Local fishermen also knew the sands and channels like the backs of their hands and often put this to good advantage. On one occasion Dennis was in his own boat fishing near another skippered by 'Slacks' Hubbard. They were about 26 miles NE of the Knoll Light and had been out two nights. Dennis took up the story, "....We got about 140 cran of herring aboard and had shot away again Saturday night when it came on a strong east wind. He came and spoke to us and we were half hauled so we had half a fleet of nets to haul when he went past us. After we'd finished hauling we had a fair wind at Smith's Knoll and then all of a sudden the wind came SSW with an ebb tide. We were five or six miles below the knoll. Then I said to our mate, 'It's no good us a punching up there with this ebbtide, we'll have to cross over there to the North Float and when the ebb's finished we'll take the flood up through the Wold,' which we did, and we were up to Yarmouth and landed half our herring before Slacks Hubbard got there although he left before us. I said, 'We gave him the sailing ship touch this morning'."

The advent of the steam drifter brought the English fishermen into conflict with Scottish fishing traditions and lead to trouble. Dennis George recalled, "....We came ashore in Scalloway, the north west part of Shetland, and there was a riot on. We took our sample ashore (a sample of their catch for the salesmen) and they got kicked overboard. They knocked the people about with truncheons, the blood flowed everywhere. They didn't want us to fish until the crofters started fishing. A liner (line fishing) brought up in the bay and we went and sold our fish to him for bait."

In 1905 Dennis George took part in efforts to find new grounds for a rapidly expanding herring industry. "....We went to try and find herring in the winter time. Three of us went searching for herring. We went ninety miles from Dennis Head, that's the NW part of the Orkneys, and at 97 miles NW went into 600 fathoms of water and shot the first night. If we'd gone another ten miles we'd have been in 1000 fathoms. We got great big herring sixteen inches long, about eleven cran. We found plenty of herring ENE of the Skerries, a cluster of rocks where a lot of trawlers used to get lost coming from Iceland. They have a light on it now, but there was no light on it then. That lay about thirty miles from Scalloway. We found plenty of herring but couldn't get any money for them.

"We had orders from Lerwick to come to Aberdeen, take our nets out and get a fresh fleet of nets in. Our store was at Aberdeen, and then we were to go up the Moray Firth through the Caledonian Canal and across to Ireland and fish out of Killeybegs and Sheepshaven. I was in the *Foxglove* steam drifter (one of Westmacott's fleet: *Sweet Briar, Honeysuckle, Heather, Bracken* etc.) When we got to Ireland we couldn't catch any herring, the biggest shot was four or five cran. They paid £3 a cran but we couldn't catch any. We stopped there a fortnight.

Waiting to go through the locks of the Caledonian canal at the end of Loch Ness. These Yarmouth drifters are typical of the vessels in which Dennis George made the same journey.
PHOTO: Mr. G. Wacey

"There was only one coal hulk and there were three of us out of the company while the other seven went to the westward after mackerel. When we first went alongside the coal hulk he gave us seven baskets to the ton, next time it was five, next time it was only four. We said, 'What's wrong? ' He replied, 'That's all the coal we've got in. There was no more to be had.' Then we got orders to leave there and finish fishing at Castle Bay in the Hebrides which we did. We came up to Scalloway back of the Orkneys, up the Longhope and coaled there, and from there through Pentland Firth to Aberdeen."

What a wonderful voyage. Through the Caledonian Canal, including Loch Ness and the Loch Gates right down to Fort William, Loch Linnhe and across to Ireland. "We were away twelve weeks," Dennis told me, "and we steamed 3600 miles!"

The last photograph of Dennis George.
PHOTO: The George family.